The Stadium in the Sky

Table of Contents

Author's Note

The Stadium in the Sky is the second part to Welcome to All-Stars Stadium. Some of the concepts in the first book have been repeated, but many have not. It is better to read the first booklet first. This second booklet does come with a photographic star chart which allows one to directly view the entire area of observation without reading the first booklet. Regardless, we definitely recommend readers to take the initiative and take their own direct measurements.

Introductory Quotations

"Mathematics is the language in which God
created the Universe." - Galileo Galilei 1564-1642

"In the sciences, the authority of thousands of opinions
is not worth as much as one tiny spark of reason
in an individual man."
- Galileo Galilei 1564-1642

Chapter 1

The World Hasn't Seen It!

"All truths are easy to understand once they are discovered; the point is to discover them." - Galileo Galilei

If one studies a little about the northern stars, the stars in the 'Big Dipper' are used as starting points for 'lines' that point to many locations further off in the celestial sky. The most well-known of these is the 'line' that one can make which points towards the North Star from the two cup end stars. This is a useful guideline, but like the other starting point lines, they only point in the general direction of their targets.

Despite all these general and inaccurate linear assertions, *there is never any mention of the two very straight three-point lines found entirely within the 'Big Dipper'.* We feel that these two three-point star lines both appearing as straight from our vantage point are convincing enough by themselves to propose that they did not happen by random occurrences. With *ten more connected three-point star lines* in our starting 'lineup', however, we don't need to emphasize that.

We do want to emphasize something else at this time. The fact that one of the three-point lines in the 'Big Dipper' is comprised of the three brightest stars in the entire northern area of observation and has still gone unrecognized is *what is definitely most amazing to us.*

Many of the remaining ten three-point star lines are also made up of true 'All-Stars', including the North Star, the 'Little Dipper' and the 'Big W'. They are all circumpolar northern stars that can be easily seen on any starry night all year. All the stars in the 'star grid' except for the home plate star are the first to be seen every night in the northern area. In light of all this, *we have every reason to expect the world to wake up soon.*

Chapter 2

Twenty to Four and We Want More

**"By denying scientific principles,
one may maintain any paradox." - Galileo Galilei**

The twelve three-point lines are all connected at both ends with only four exceptions, and these four lines are also connected to the star grid but only at one end. *Twelve three-point lines connected at 20 of 24 end points with none disconnected* - there has to be a formula for describing just how remote the chances are for just this aspect of the grid being caused by random occurrences.

We have already stated that if all twelve three-point lines were disconnected from each other, they would still by themselves make all too strong a case for their 'structural validity'. A first primary formula or equation exists that would calculate the remote chances of these twelve three-point lines having occurred by many separate coincidences, *without factoring in any of the connecting points. It might seem like one was going too far if an attempt was made to describe just how remote the chances might be for both of the above formulas to have simultaneously occurred by random occurrences. That is exactly how the grid is put together, however, and so any attempt at calculating the overall remote chances for the grid being formed by coincidences should definitely include both formulas, as they simply both exist.*

There are many other factors that still exist including the relationships in magnitude that we have previously described. These other factors do also exist and so they should be included in any overall formula(s) or equation(s) for stating the extremely remote chances or odds of the entire star grid having occurred by random occurrences. We will go over some of the many remaining factors to the overall equation shortly.

With the score already 20-4, it really doesn't matter but we gotta tell ya' something baseball fans. Two of the four so-called disconnected end points or non-connecting point end point stars are *second base and home plate!!* If second base and home plate aren't connecting points, you gotta be from another galaxy or something... so it's 22-2 fans, not that we need the insurance or anything. But g'wan, go away with that three-point stuff for just a minute here. The rest of the constellations are all just two point pointless figments of somebody's imagination, right? What, the base lines aren't good enough lines for you? *Ahh, its OK fans, we got it all wrapped up.*

Chapter 3

Observation Point Earth

"It is surely harmful to souls to make it a heresy to believe what is proved." - Galileo Galilei

The distances between the stars that belong to the star grid are certainly great, yet the overall grid is an enclosed and confined pattern that clearly resembles familiar 'blueprints' or 'layouts' found in various forms of construction, including our example of mountain property or mountain fence lines. Rather than some far-fetched or far-stretched bunches of lines extending further and further from the observer (or observation point) the altogether tightly closed hexagon exterior to the pattern allows for a single observation point to be used in a very common manner with that of many construction techniques. The remaining six lines also fall almost completely *within* the hexagon. Other than a 'small' part of the three lines that connect into the 'Clipper' star, which is in the left center field seats, just one of the lines extends 'slightly' beyond the six-sided diamond at one end point. The confined nature of these remaining six lines indicates their 'construction' would also be attainable from the very same observation point. In relationship to each of the twelve three-point lines, this is a fourth mathematical point that must be included in the overall equation.

Once again, the 'down to Earth' example of mountain fence line construction clarifies the above. If a skilled mountain fence builder or surveyor were to stand on the side of a canyon or mountain slope and direct another builder who was located on the other side of the canyon, they would have little difficulty (using just markers instead of fence post) establishing a six-sided diamond of connected three-point lines along with the six interior three-point lines. In fact, the surveyors could quite easily establish the same enclosed pattern as found in the star grid. Just like the intricate star pattern, all twelve of the connected three-point lines would only appear as straight when viewed from the observation point. In the case of the star grid, this single viewpoint is Observation Point Earth.

Chapter 4

Facts, Facets, Factors and Formulas

"Facts which at first seem improbable will, even on scant explanation, drop the cloak which has hidden them and stand forth in naked and simple beauty." - Galileo Galilei

The existence of twelve different three-point star lines is highly unlikely to have occurred by random occurrences, but it is only one factor involved in the overall star configuration from which a separate formula can be derived. The fact that all of these twelve three-point lines exist in the same northern area is another *separate* factor. The fact that all the stars of significant magnitude in the area of observation are included in the star grid *without one remaining exception* is another *separate* factor from which a separate formula could be derived, and both of these factors also greatly increase the chances of this overall situation having occurred by many random occurrences.

The fact that just two or three of these three-point lines should be connected at even one end is highly *unlikely*. As we noted in Chapter 2, the fact that no less than twenty of the twenty four end points to the twelve three-point lines are connected represents another *separate factor and very large separate formula* that greatly increases the chances of the entire star grid being just a product of many coincidences. The fact that there are three triple-connecting points increases the chances even further.

The fact that the entire intricate formation has only two examples of three-point lines that cross over each other without the presence of a connecting point star is yet another factor that increase these odds even further. The fact that the three brightest stars in the entire area of observation belong to the same three-point line and the fact that these three stars have virtually the same magnitude are two factors that just by themselves alone are highly unlikely to have happened by random occurrences. The fact that three of the four 'Grand Slam Lines' appear to run so much in the 'same direction' across the night sky that together they form an elongated arrow shape is another separate factor, and the fact that these three 'close' lines are actually separated at times by hundreds of light years in depth is another very separate multi-faceted factor. The fact that the path of the other 'Grand Slam Line' should intersect the *only* other star of relatively the same magnitude within the entire area of observation (the other very bright star in the 'Little Dipper') as the one that is the triple connecting point for the other three lines (the North Star) is yet another separate factor from which a formula could be derived. The fact that another three-point star line should use this same star along with the only other star of even close to that magnitude in the entire area of observation (the other fairly bright star in the 'Little Dipper') and then connect into the peripheral six-sided diamond at the other end point represents a combination of facets that by themselves are highly unlikely to have occurred by random occurrences. The shape of the enclosed 'six-sided diamond' and the remaining six lines that are virtually enclosed

within this hexagon is another separate factor and there are
many more including all our observations on the consistent
distribution of magnitude from the 'core' handle throughout the
star grid.

It is extremely interesting that after all 21 of the
connected stars are accounted for, there are virtually no other
stars of *even lesser but still significant magnitude whatsoever*
within the entire area of observation. There is one star positioned
in fairly straight away right field. This star (which is part of
Cepheus) is at least brighter than anything else within the
considerable confines of All-Stars Stadium. Once this quiet star
makes a reluctant appearance, it can clearly be seen above all the
rest, and so we have named this star right fielder the 'Rajah'.

The formulas or theories that can be derived from these
separate factors also stand separately from each other. Each factor
and formula states a completely different reason from the next as
to how the star grid could not likely have happened by many
random occurrences. Given all the factors and formulas that each
point to a 'zero' chance of the star grid having occurred by many
random coincidences, one must ask at what point do all these
combined formulas so outweigh any likelihood of the entire grid
having occurred by random occurrences that the 'critics' of our
observations have to literally give a little in terms of the accuracy
of some of the twelve connected lines? We believe the 'critics' can
be shown some imperfections and, like a team facing immediate
mathematical elimination from the playoffs with still two weeks
left on the schedule, they can try hard but they will still have to

accept the reality of the overall star grid's very solid 'structural validity'.

We are not suggesting that the twelve three-point lines are not accurate. We are very confident in their accuracy and we will discuss this next.

Chapter 5

Taking Correct Measurements

"Measure what can be measured; and make measurable what cannot be measured." - Galileo Galilei

We don't really believe in the necessity of taking sophisticated measurements with the most advanced technology on the accuracy of the connected three-point lines. We do realize that a skeptical public will want this information and it is very understandable that the people should be given these stats. We want these stats to be based on the correct data and we do have a few things we would like to point out.

We would expect that the measurements will be taken directly from the area of observation and no photos will be used. There are some photos that show all twelve of the three-point lines to be straight, but there are many that don't. As we have mentioned before, photographic accuracy is something that is often very unreliable with stellar subjects, especially in regard to accuracy of location. This is so well understood that we would not even mention it, but we do have a concept that we want to be well understood and we can use photographic evidence as a first example.

A photograph of a three-point star line transfers that image to a linear plane piece of photo paper. Using the same image, the apparent width of the third 'target' star (the approximate real width of each of these stars is already available) can then be measured against the apparent length of the line. A direct ratio between the two can then be determined. If technology can determine the same from direct measurements, then the use of photographs won't be necessary. *It is absolutely necessary to establish this preliminary apparent linear distance / apparent width of target star ratio.*

Our down to earth fence model provides an excellent example of this ratio. When a mountain fence builder guarantees that a quarter mile section of wire fence will be within one quarter inch of perfectly straight, the apparent linear distance (and in this case the approximate real linear distance) over the width of target tolerance ratio is 63,360 to one. Although this ratio may sound very specific or accurate, a quarter inch of tolerance for a quarter mile would be very inaccurate work by a mountain surveyor's standards. The considerable dimensions of All-Stars Stadium offer the possibility of much more exact ratios to be calculated, indicating the kind of precision that even surveyors would envy. What remains to be figured is the amount of inaccuracy in each of the twelve three-point lines. We believe that if only the apparent linear distance is used in measuring the lines as opposed to the real linear distance, the amount of inaccuracy derived from such a calculation is too great.

We definitely agree that preliminary measurements need to be taken directly from the area of observation. *We also believe there are at least two very big reasons why direct measurements by themselves do not indicate an accurate full linear length / target width tolerance ratio and therefore are misleading.*

In order to attain a truly accurate measurement on any of the three-point star lines a preliminary measurement on the percent of angle inaccuracy has to be taken first using the ratio derived from the apparent linear distance / apparent width of the star (amount of target tolerance). *The amount of recorded inaccuracy from this direct measurement then needs to be adjusted to accurately reflect two very large factors that need to be accounted for.*

The first of these two factors is that the star grid is three dimensional and therefore direct measurements only view the apparent length of each line and not the real length. The real length is always a considerable distance more, and with some stars located much further away than others, the difference in distance is often vast.

The second of these factors may have an even greater effect on the linear length / target width tolerance ratio of these lines. *The distance down the length of the three-point star lines is measured in light years, whereas the width of the star (even with greater apparent width from light refraction included) is measured in miles. The distance down the length of these three-point star lines is so vast (some are hundreds of light years long) that the apparent distance from a direct measurement simply does not accurately show this great length when directly compared to the measurement of the width of the target star.* * Measurements previously adjusted to reflect the difference in dimension still do not reflect this.

Once factors like the width of the target star are
established as best as possible, the apparent measured distance
of the line then can also be calculated as best as possible.
Using the width of the target star as a constant, the apparent
measured distance ratio can then be directly compared to the real
distance ratio with a final direct ratio between the two then being
determined. This same final ratio then needs to be applied to the
apparent percentage of inaccuracy that had been recorded by
direct measurement, resulting in a real percentage of inaccuracy.

The real distances to the lines in All-Stars Stadium are much greater than the apparent distances, and so the real percentage of inaccuracy in these lines is proportionately less than the apparent measured inaccuracy.

The establishment of a constant ratio to ratio differential that would apply to all the three-point lines will not work of course. It will be necessary to establish a separate specific ratio for each three-point line. We would expect the experts to have little difficulty with all of this and we would hope they would know what needs to be done. In summary, we are basically stating that the recorded amount of measured inaccuracy on each of the twelve three-point star lines has to be adjusted from the direct apparent measurement based on the apparent linear distance to one based on the real linear distance. We still believe in our direct traditional method of measurement, but we also agree that in today's world the need to establish modern scientific measurements as best as possible is real. If the 'ground rules' we have stipulated are included in the final readings, we will be

more inclined to accept these readings.

*The apparent distances taken from direct measurements also
definitely do not indicate the often large differences in real linear
distances of three-point star lines that have relatively close
apparent distances. The three 'Grand Slam Lines' that share the
North Star as their middle connecting point provide a truly
excellent example. Each one of the three lines begins at one of the
stars in the handle of the 'Big Dipper.' These three stars not only
appear as 'close' together, but the apparent distances down each
line to the North Star are not that far apart. Two of these handle
stars, the left-left center field star (Mizar) and the 'Bullpen Gate'
(Alioth) are actually very close in terms of their distance from
Earth. Mizar is 82.8 light years away from Earth and Alioth is
82.6! The third star, the 'Clipper' (Alkaid, which is not a member
of the Ursa Major Moving Group), is 104 light years away from
Earth. The path these three stars take to the North Star is not that
different in apparent linear distance as the North Star is now
estimated to be about 350 light years away from Earth. Both
apparent linear distances are remarkably close to each other (as
are the real linear distances) for the two lines running from
Mizar and Alioth until the two lines reach the North Star. After
the North Star triple connecting point, the three lines take a much
more separate route from each other, although the apparent paths
and the apparent distances of the three lines appear as similar as
ever from our observation point. The line from Alioth (the
'Bullpen Gate') through the North Star is 550 light years from
Earth when it connects into Cassiopeia's second base star
whereas the line from Mizar (the left-left center field star)
through the North Star is 99 light years away when it connects
into third base. The third line from Alkaid (the 'Clipper') appears
to share a similar route as the other two, but it is 410 light years
away from Earth when it connects into the left infield dirt star in
Cassiopeia! Despite the undeniably similar paths and apparent
distances that these three lines make in forming the very artistic
'Northern Arrow', there is a vast difference between the real
distances of these three three-point star lines beginning from the
North Star and onward into the three 'target' stars in Cassiopeia.*

24

and the
'NORTHERN ARROW'

Using a straight edge,
please verify
the following twelve
connected
three-point lines:

1,2,3
4,5,6
7,8,9
10,11,12
13,14,15
16,17,18
19,20,21
ABC
DEF
GHI
JKL
MNO

Chapter 6

The Jolt in Joe's System

"In time you may discover everything that can be discovered, and still your progress will only be progress away from humanity. The distance between you and them can one day become so great that your joyous cry over some new gain could only be answered by a universal shriek of horror."
- Galileo Galilei

Chapter 8 in Welcome to All-Stars Stadium, A Question of Magnitude, reveals a second different relationship between the same twelve three-point star lines in the northern night. Another overall summary of the relationship in magnitude within the star grid should provide a more complete explanation of how the light in 'the system' is consistently distributed.

The three stars in the handle of the 'Big Clipper' (the 'Big Dipper') have the greatest amount of connecting points (five) of any three consecutive 'close' stars in the grid, and these same three stars are three of the brightest in magnitude in the entire area of observation. *The 'Big Dipper's' cup end star that is the third star in the three-point line that runs through the two brightest handle stars is almost the exact same brightness as these two.* The first primary consideration is that the three consecutive stars that have the most connecting points between them are three of the brightest in magnitude (along with the only star that actually has a 'double connection' to the handle and the also directly connected North Star) in the entire area of observation.

As we have noted previously, one could include the North Star as part of the system's 'core' along with the handle of the 'Big Clipper'. We prefer to designate the handle by itself as the three stars appear and in fact are much closer and together they form the tail to the 'Northern Arrow' that indicates the start of the 'jolt's' direction to us.* The North Star is not quite as bright as two of them either.

On the other hand, the apparent magnitude of the North Star (1.98 Apparent Magnitude) is slightly brighter than one of the three stars in the handle. If one did include the North Star (which is actually a multi-star system) as part of the core of the light distribution, this four star group would represent a strengthened core as it now would include three triple connecting point stars in a row along with a fourth connecting point star, the handle's 'Bullpen Gate'. The center of the system would now represent seven connecting points to incoming three-point star lines, and would form a true *'Core Four' group of All-Stars*. That definitely sounds stronger to us... so either way. Regardless of how the core may be most correctly defined, ***it is truly remarkable how close the apparent magnitudes are for these four brightest stars and the 'double jolted' 'Yogi' star, all of which are directly connected (the core's 'Clipper'-1.84 AM, the left-left center star -2.23 AM, the 'Bullpen Gate'-1.76 AM, and the flanking 'Yogi'-1.79 AM and North Star-1.98AM).***

The main distribution in magnitude from the three core handle stars runs in four 'long' three-point lines that connect into four separate stars in 'Cassiopeia', the 'Big W'. This distribution runs in the same direction as the pointed arrow (with base) shape that the lines create! Three of these 'long' three-point lines have the brightest star within the entire area between the 'Big Dipper' and 'Cassiopeia', the 'Little Dipper's North Star', as their shared triple connecting point. Beginning at the 'Clipper' star at the end of the handle, the fourth three-point line runs through the next brightest star within the entire area. This star is also part of the 'Little Dipper' and is also a connecting point star in the entire grid's framework and therefore should be bright. This star is positioned in left center field close to what would be straight away for a right handed hitter considering the 'tape measure' size of 'Death Valley' out there. Rather than 'Kochab' or the 'other bright star in the 'Little Dipper' we have given this star the name the 'Magnificent Mick' or just the 'Mick'.

The distribution of light begins from the three core handle stars of greatest magnitude through the North Star (1.98 AM) and the 'Mick' (2.08AM), the other bright star in the 'Little Dipper' and these same stars that are the next two in line from the core are also the next two brightest stars in the area of observation. The triple connecting point North Star is the *slightly* brighter of the two.

The four stars in receptive 'Queen Cassiopeia' that are the four end points to the 'Grand Slam Lines' include the stars at the next descending level of brightness. With the two connecting points spread out evenly at first and third base, the magnitudes of the infield All-Stars demonstrate the consistent descending level of brightness and the balanced distribution. The 'first base star' is a connecting point from where the light in the grid then runs down the first base line, and it is the brightest star in Cassiopeia (2.24AM). The right infield dirt star (2.28AM), second base (2.47AM) and third base (2.68AM) are also very close. It is interesting to note again that the left infield dirt star is the furthest away from the first base star and is also a non-connecting point and it is much dimmer (3.37) than the other three end points (first, second and third base) to the 'Grand Slam Lines'. And yes, it's also in foul territory...and more than 400 light years away!

It is truly remarkable to note how the two stars on each side of first base appear to be virtually equal in both apparent magnitudes and apparent distances from first base when the actual differences are dramatically different. First base is 228 light years from Earth whereas the right infield dirt star is 'only' 55 light years away and second base is ten times further away at 550 light years from Earth! The intrinsic differences in brightness are as far apart as the real distances in light years, yet everything appears remarkably the same. If we appear to have become distracted by these amazing aspects of what seem to be just two-point base lines in space, it is important to not forget that each of these three stars belong to one of the three-point 'Grand Slam Lines'!

Second base is a pulsating variable star that occasionally outshines first base in apparent magnitude. In ancient times first base was known as the 'Breast' and third base was the 'Knee' of Queen Cassiopeia. It is easy to be truly distracted by the irresistible Queen and her powers that draw the 'Four Grand Slam Lines' right to her. With her gleaming belt rising like the pulse of the admirers of Cassiopeia, the Queen's star sequined dress shimmers across the Sexlectric Universe.

The distribution of light continues from the brightest star in Cassiopeia, the connecting point first base star (2.24 AM), down the first base line to the right field foul pole star (which is the brightest star in the constellation 'Cepheus'). The right infield dirt star (2.28AM) is a non-connecting point on this first base line. Its location appears to be much closer to first base than the right field foul pole at the far end. The apparent magnitude of the right infield dirt star is less than the preceding first base star, but it is brighter than the following right field foul pole connecting point. We believe this straight one-two-three lowering of light distribution can be explained by the proximity of the first two stars in comparison to the relative middle location of most of the other non-connecting point stars in the star grid. Most importantly, the consistent declining distribution of light continues down the first base line.

The right field foul pole star (2.45AM) continues to display the proportionately less amount of light that is being distributed down the lines, and as a connecting point to the right field wall it is brighter than both of the stars in the right field wall that follow in the path of distribution. The non-connecting point right field wall star *is* the least bright of the three (Nodus Secundus 3.07AM); although the right-right center star (Zeta Draconis 3.17 AM) is listed as being slightly less bright, and it is quite remarkable how consistently proportionate the descending distribution of light remains this far down the lines. Like the batting order in the All-Star Game, the right field wall stars may be less known than the big stars in the 'line up' before them, but these are the ones that convincingly keep it going.

The Yankees moved into their new stadium in 2009 and won the world championship that year. Members of the team said the winning 'Bronx Magic' had moved into the new ballpark from the old Stadium across the street. Some said that some of the magic went from the old right field wall into the new one. All-Stars Stadium is like both Yankee Stadiums, the 'House That Ruth Built' and the 'House That Jeter Built', and it set the precedent with a tantalizing right field porch.

The next star down the lines, the right center field wall (Eta Draconis 2.73AM) is brighter than the previous right-right center connecting point star (3.17AM), but it *is* a connecting point with the three-point star line that runs from the two bright stars at the cup end of the 'Little Dipper'. This three-point star line (which we sometimes refer to as the 'bat rack') provides a shortcut for the 'jolt' originating from the core handle and so the right center field wall star is somewhat brighter than the preceding right-right center field wall star. As usual, the middle non-connecting point star in the 'bat rack' (Pherkad 3.05AM) is a little less bright than the two connecting point stars at each end, and the connecting point star nearer the beginning of the direction of the light distribution is the brightest.

The last star in the right center field wall, the center field star, is actually a third connecting point in a row and it is the last of the three to receive the light distribution. The center field star (Theta Draconis 4.11 AM) is also the very last star at the end of the path of light distribution, and it is the dimmest of all the stars in the entire preceding 'lineup'. The next star, the left center field star, has a direct connection back to the core handle. We will discuss all the remaining stars within the grid that have direct connections to the handle in the following chapter.

* As per Welcome to All-Stars Stadium, we fully support the most widely accepted contention that starlight is a result of burning forces within each star. We only want to demonstrate the pure mathematical principles that are evident throughout the interstellar relationship of the star grid.

Chapter 7

Joltin' Joe's System

"My dear Kepler, what would you say of the learned here, who have steadfastly refused to cast a glance through the telescope? What shall we make of this? Shall we laugh or shall we cry?" - Galileo Galilei

There is also some definite distribution of light from the three bright stars in the core handle that is directed out in all three of the remaining three-point lines that are also directly connected to the stars in the handle. First, there is the line within the 'Big Clipper' that runs through two of the very bright handle stars. This 'double jolted' line ends with its third star, the star at the top lip of the 'Dipper' cup, also being very bright, in fact *the next brightest star along with the first two stars on this same line in or within the entire area of observation!* In regard to this third star ending up almost exactly as bright as the first two, well a 'double jolt' would do that! We call this very bright star that is located in the left field corner near the foul pole the 'Yogi' star, in recognition of his ability to play left field. ***This brightest three-point star line has somehow gone unrecognized, but with these three equally bright stars now being comprised of no less than the 'Clipper' (1.84AM), the 'Bullpen Gate' (1.76AM), and the 'Yogi' (1.79AM), we hope to change things.***

The three-point line that is the left field wall begins at the middle binary star in the 'Big Clipper' handle (the left-left center field star or Mizar 2.23AM), and finishes at the star at the bottom far end of the 'Dipper' cup. This star (the left field foul pole star or Merak), receives one direct 'jolt' from the handle and it is also a connecting point star. Accordingly, this jolted star is also very bright (2.37AM), but not as bright as the double jolted star.

It is most interesting to note something about the middle star in this left field wall. This star is next to the three stars in the core handle, but it is a non-connecting point and it is *much* dimmer. The overall dimness of this star (the left field wall star or Megrez 3.31AM), the star where the handle 'connects' in to the Dipper cup, is another indicator that most of the light from the handle is being distributed down the 'Four Grand Slam Lines'. The direction of this distribution of light begins from the 'tail' shaped handle and runs where the connected stars in the 'Northern Arrow' point. The arrow/base shape is completed in the infield 'Cassiopeia', but, as we have described, the proportionate distribution of descending light continues on from there.

The seventh and final connected three-point star line running out from the three core stars in the handle is the line that runs from the left-left center field star in the middle of the handle and becomes the left center field wall. The next star in this line, the left center field wall star, is definitely brighter

than the following center field star. This clearly indicates
that at least some 'jolt' from the core handle middle star is also
distributed down this seventh three-point star line. *The case
that all seven three-point star lines running from the 'Big
Clipper' core handle receive consistent distribution of
descending brightness is very valid.*

*As we have described, the proportionate distribution
from greater to lesser magnitude (and with connecting point
stars brighter than non-connecting point stars) continues from
where one of the 'Four Grand Slam Lines' connects into the
brightest star in 'Cassiopeia', the first base star. The remarkably
consistent distribution of light continues from there in an
unbroken chain down the first base line and around the outfield
walls of magnificent All-Stars Stadium.*

We have tried to describe the consistent distribution of
magnitude within the star grid. This relationship is yet another
part of the overall structure that is the All-Stars
Stadium/Northern Arrow star grid of our northern sky. Our
primary model remains the twelve connected three-point star
lines. We have demonstrated the structural validity of these
connected lines, and we have also pointed out the relative
simplicity that is apparent in both this model and its
irrefutability.

Our description of the relationship in magnitude among
the stars in the northern area of observation involves a
completely different intricate relationship between these same

stars than the primary contention of twelve connected three-point lines. Just as the relationship in magnitude involves a different intricate pattern than the connected star lines, the chances of the overall combination of these patterns having been caused by many random occurrences are even less likely than the already very remote chances for the twelve connected lines.

So how remote are these chances? Well, baseball fans, it's like the big ballpark. We're talkin' light years... the chances are so remote, they're... astronomical! It's outa the park fans!

Chapter 8

A Question of Attitude

"Long experience has taught me this about the status of mankind with regard to matters requiring thought: the less people know and understand about them, the more positively they attempt to argue concerning them, while on the other hand to know and understand a multitude of things renders men cautious in passing judgment upon anything new."

- Galileo Galilei

"Very great is the number of the stupid." - Galileo Galilei

We have mentioned how the star grid of connected three-point star lines appears to be aimed at us, as it is only from Earth's region that all of these lines appear as straight. After considerable time is spent directly observing the area of observation, it is remarkable how the star grid that we speak of also *seems* to be aimed at us. Aside from the extreme unlikelihood of having been somehow created by dozens of separate coincidences, all the stars of significant brightness in the vast area being included in the pattern plays a supporting role to the *overall undeniable grid-like appearance* of the northern area of observation.

This structural grid like appearance that we speak of is then exemplified by formidable factors like the six sided diamond and the containment role it holds on the other six lines, 'locked in' lines like the stabilizer 'bat rack', the balanced appearance of five separate 'twin grid' two star formations (already a part of some of the three-point lines) that point close to the same direction etc. These 'twin grid' stars include the two stars made up of the right field foul pole star and the fairly bright star that is in the seats directly behind the right field foul pole and also close to the (extended) first base line. We wrote previously that alterations like those done in New York after the 1923 season were apparent on the first base line. The existence of this star behind the right field foul pole may mean that it is better positioned for an accurate three point line between it and the right infield dirt star and first base. The truth is, we do not have the same confidence in accuracy for the first base line as we do for the other eleven three-point lines, but the recognition of this 'twin grid' star behind the right field foul pole star may definitely result in a much more accurate three-point line being established down that first base line.

Just as *familiarity brings greater recognition* in the above sequence, we clearly believe that the wrong preliminary attitude can become a very limiting factor in assessing the star grid. In regard to the twelve connected three-point star lines, we believe that once one develops an attitude of acceptance to the grid's structural validity, one is no longer limited in making real assessments of it.

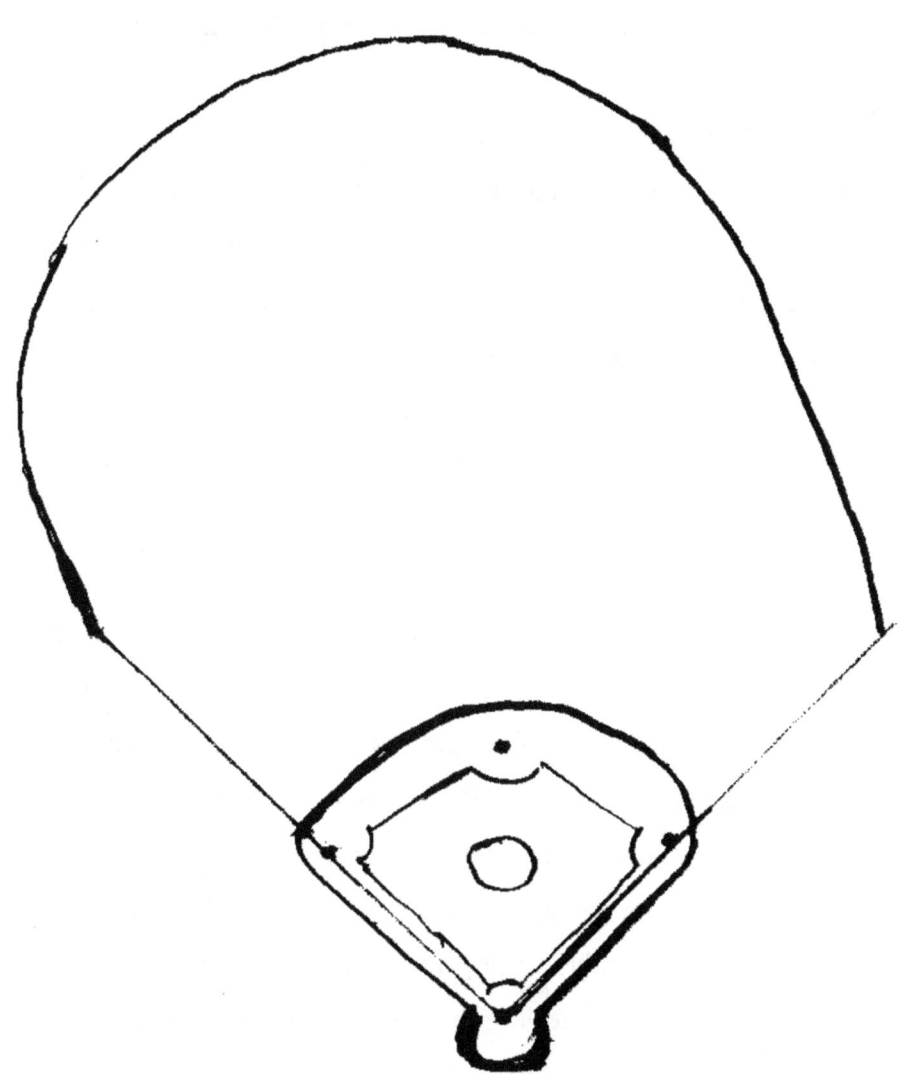

YANKEE STADIUM 1923

Chapter 9

The Role of Baseball and the Home Plate Star

"It ain't over 'til it's over." - Yogi Berra 1925 -

A few of the stars in the star grid are moving in entirely different directions than the rest. Most if not all of the three-point star lines were probably viewable by ancient cultures. In the past, these connected star lines did not receive real recognition, and all of these star lines won't last forever either. The six-sided diamond known as All-Stars Stadium has been spotted now and this is the Age of Baseball. We are the ones who spotted the message in the northern sky, and we're baseball fans. It seems logical to us that some of this must be about baseball.

We pointed out previously how the annual precession of All-Stars Stadium happens to have the right field line on the right side and left field on the left starting at the beginning of the baseball season and finishes with the most convenient time for viewing being right at the end of the season!

There are other separate factors that demonstrate the role of baseball in the northern star grid. First, if one separately analyzes the overall pattern of the four outfield walls from foul pole to foul pole, the similarity with many of today's ballparks is remarkable. Secondly, once the left infield dirt star is understood to be in foul territory where the corner of the dirt touches grass, the 'Big W' looks more like an overhead view of an infield.

This resemblance is then clearly strengthened by the mystical appearance of the legendary home plate star.

The home plate star has the dimmest magnitude in All-Stars Stadium and it should. First, the turn in the first base line inside of first base indicates that home plate is *not* a real connecting point star, so according to our observations on magnitude it should be of somewhat lesser magnitude to begin with. The fact that the home plate star is located underneath the 'W' in Cassiopeia definitely indicates that it is also *behind* the four stars in Cassiopeia that receive the 'jolt' coming down the four 'Grand Slam Lines' from the core handle. The home plate star is the least bright star in the ballpark, as it is fully protected by the infield shield of the four stars.

With the infield cutting off all the action, it seems like there's nothin' for a Yogi to do back there. Home plate is also the least visible star in the star grid by far, and it is completely unneeded in determining such an intricate grid's structural validity.

Home plate does exist, however, and it is right where baseball fans want it to be. Its existence establishes the plate and completes the last line in the baseball diamond, the third base line. All-Stars Stadium is now complete and the beauty of baseball is the final and ultimate message from the northern night. Berra makes the tag and it's over!! How'bout dat!!

www.ingramcontent.com/pod-product-compliance
Lightning Source LLC
Chambersburg PA
CBHW070721180526
45167CB00004B/1574